CW01510008

Net Technical Assessment

A Methodology for Assessing Military Technology Competition

JON SCHMID, CHAD J. R. OHLANDT, SHAWN COCHRAN

Prepared for the Office of the Secretary of Defense
Approved for public release; distribution is unlimited

NATIONAL DEFENSE RESEARCH INSTITUTE

For more information on this publication, visit **www.rand.org/t/RRA1350-1**.

About RAND

The RAND Corporation is a research organization that develops solutions to public policy challenges to help make communities throughout the world safer and more secure, healthier and more prosperous. RAND is nonprofit, nonpartisan, and committed to the public interest. To learn more about RAND, visit www.rand.org.

Research Integrity

Our mission to help improve policy and decisionmaking through research and analysis is enabled through our core values of quality and objectivity and our unwavering commitment to the highest level of integrity and ethical behavior. To help ensure our research and analysis are rigorous, objective, and nonpartisan, we subject our research publications to a robust and exacting quality-assurance process; avoid both the appearance and reality of financial and other conflicts of interest through staff training, project screening, and a policy of mandatory disclosure; and pursue transparency in our research engagements through our commitment to the open publication of our research findings and recommendations, disclosure of the source of funding of published research, and policies to ensure intellectual independence. For more information, visit www.rand.org/about/principles.

RAND's publications do not necessarily reflect the opinions of its research clients and sponsors.

Published by the RAND Corporation, Santa Monica, Calif.
© 2024 RAND Corporation
RAND® is a registered trademark.

Library of Congress Cataloging-in-Publication Data is available for this publication.
ISBN: 978-1-9774-1292-8

Cover photos, top row: Bernardo Fuller/U.S. Army, Gary Singleton/U.S. Army, Richard Eldridge/ U.S. Air Force; bottom row: U.S. Space Force, Samantha Rodriguez/U.S. Marine Corps, Warren Duffie/U.S. Navy.

Limited Print and Electronic Distribution Rights

This publication and trademark(s) contained herein are protected by law. This representation of RAND intellectual property is provided for noncommercial use only. Unauthorized posting of this publication online is prohibited; linking directly to its webpage on rand.org is encouraged. Permission is required from RAND to reproduce, or reuse in another form, any of its research products for commercial purposes. For information on reprint and reuse permissions, please visit www.rand.org/pubs/permissions.

About This Report

Technology development is a recognized front on which strategic competition between the United States and other nations is contested. Net technical assessment (NTA) constitutes a means of assessing the state of play in competition for a given technology or technological domain. The Office of Strategic Intelligence and Analysis (OSI&A) in the Office of the Under Secretary of Defense for Research and Engineering (OUSD[R&E]) is conducting NTAs for U.S. Department of Defense (DoD) modernization priorities to assess the relative strengths, weaknesses, opportunities, and threats that the United States faces in the conceptualization, development, and deployment of specific technologies within DoD modernization priorities. In support of that effort, OUSD(R&E) asked RAND to develop and execute an NTA methodology. This report describes that methodology.

The research reported here was completed in December 2023 and underwent security review with the sponsor and the Defense Office of Prepublication and Security Review before public release.

RAND National Security Research Division

This research was sponsored by OUSD(R&E) and conducted within the Acquisition and Technology Policy Center of the RAND National Security Research Division (NSRD), which operates the National Defense Research Institute (NDRI), a federally funded research and development center sponsored by the Office of the Secretary of Defense, the Joint Staff, the Unified Combatant Commands, the Navy, the Marine Corps, the defense agencies, and the defense intelligence enterprise.

For more information on the RAND Acquisition and Technology Policy Center, see www.rand.org/nsrd/atp or contact the director (contact information is provided on the webpage).

Acknowledgments

We would like to thank OSI&A, which conceptualized, spearheaded, and funded the NTA effort. RAND researchers Julia Brackup, Jennifer Brookes,

Christian Curriden, Silas Dustin, and Sarah Harting were indispensable to the project. Technical expertise provided by the MITRE Corporation, MIT Lincoln Laboratory, Johns Hopkins University Applied Physics Laboratory, the Institute for Defense Analyses, and Georgia Tech Research Institute was invaluable to this effort. Finally, we would also like to thank the dozens of workshop participants who lent their expertise to this project.

Summary

The Biden administration's 2022 National Security Strategy recognizes technology as a major front on which strategic competition is waged, observing that "technology is central to today's geopolitical competition and to the future of our national security, economy and democracy."[1] The centrality of technology to the conduct and outcome of strategic competition motivates an analytic task: the development of methodologies for assessing the state of play of technological competition. This report proposes one such methodology for the U.S. Department of Defense (DoD): a net technical assessment (NTA), which, for a given technology area, is designed to characterize the technical state of the art, measure relative national standing, and identify and assess technology applications for accomplishing strategically important military objectives.

Our NTA methodology is informed by the long-standing net assessment approach but tailored to assess the impact of select technologies on strategic competition. The practice of net assessment was largely developed by Andrew Marshall during his time at the RAND Corporation and the Office of Net Assessment. Marshall proposed that to correctly understand a country's strategic behavior requires abandoning traditional (rational actor-based) approaches to evaluating military power in favor of a novel approach built on the detailed empirical study of the myriad factors that affect national strategic behavior in practice. Our approach leverages Marshall's insight and proposes an assessment methodology that connects demonstrated national technical capabilities in the context of scenario-based reasoning (cognizant of the complexity of competitors' domestic bureaucracies and institutions below the national level) and draws on tools, ideas, and concepts from diverse disciplines based on their practical utility in advancing the assessment task at hand.

Rather than attempting to estimate the pace and character of technological change for all technologies and assessing which of these technologies will be most important in a given future competition scenario, our NTA

---

[1] White House, *National Security Strategy*, October 2022, p. 32.

approach begins by scoping the assessment task to a technology area that is currently prioritized by both the United States and a competitor. With the critical technology area selected, we then document the state of the art and technological trends, which can then be applied to strategic challenges facing DoD to identify promising opportunities that DoD might consider pursuing.

The NTA methodology described here entails the following three research tasks:

1. assessment of the technological state of the art and trends
2. exploration of military operational implications
3. reflection on key competitor perspectives.

The objectives and deliverables associated with these tasks are summarized in Table S.1.

First, during the technology trends task, for a given technological domain (e.g., quantum science), we scoped the technologies, characterized the technical state of the art, and measured relative national standing. To assess the technical state of the art, each technology was investigated by subject-matter experts to identify the current maturity and rate of development of critical technology elements (i.e., technical inputs to an integrated or final technology). In establishing the technical state of the art, subject-matter experts also identified likely applications of the focal technologies, placing particular emphasis on those with military relevance. Supplementing subject-matter expert opinions on relative national standing, we conducted an extensive country-level *bibliometric analysis* (i.e., analysis of patents and scientific publications) and a qualitative assessment of the domestic research and development and commercial ecosystems supporting the focal technological domain.

The second major research task was to explore the operational implications of the focal technology area via a structured analytic workshop: a full-day event attended by experienced military operators, strategists, regional experts, technical experts, and others. The central analytical function of this task was the identification and assessment of technology opportunities to understand the operational relevance, technology timeline, and uncertainties associated with applications of the focal technologies. The output

TABLE S.1

Summary of Net Technical Assessment Methodology

Research Tasks: Approach	Objectives	Deliverables
Technology trends task: technical level-setting	• Scope a set of technologies • Assess technologies' technical state of the art and trends • Capture relative national standing for focal technologies	• Technical level-setting report • Technical level-setting brief • Relative national standing brief
Operational implications task: structured analytic workshop	• Identify and refine scenario-specific technology opportunities for focal technologies • Assess operational advantage and fielding timeline for each technology opportunity	• Technology opportunity list assessed for operational advantage and fielding timeline • Workshop notes (e.g., description of discussion, notable points of expert divergence and convergence)
Competitor perspective task: intelligence community engagement	• Refine all prior assessments based on intelligence	• Refined assessments of technical level-setting, national standing, and technology opportunities
Summary of full NTA process	• Evaluate the expected impact of a technology on strategic competition	• Final NTA report for each critical technology area

of this task was a set of operationally relevant technology opportunities scored by workshop participants based on expected operational impact and development and fielding timeline. The workshops also allowed us to capture technology opportunities, the associated reasoning regarding expected operational impacts or fielding timelines, and the uncertainty of the reasoning based on consensus or divergence among participants.

The final NTA research task was the integration of competitor perspectives through intelligence community engagement. The purpose of this task was to ensure that the intent and activities of potential adversaries are adequately captured in the assessment. During the intelligence community engagement, particular attention was paid to the state of competitor technological advancement, their military strategy and posture, and the existence or the planned acquisition of counters to the technology opportunities that were identified during the workshop.

The principal result of the NTA methodology is an integrated NTA document for the focal technology area that can be leveraged and built on by the defense and intelligence communities. For this effort, no specific policy questions are asked or answered, but the expanded understanding of the technologies and their role in strategic competition helps inform DoD resource allocation to both technology development and intelligence collection priorities.

Contents

Figures and Tables

Figures

Tables

CHAPTER 1

Introduction

The 2022 National Security Strategy recognizes that the United States remains in a state of strategic competition, calling out Russia as "profoundly dangerous" and China as "the only competitor with both the intent to reshape the international order and, increasingly, the economic, diplomatic, military, and technological power to do it."[1] It also recognizes technology as a major front on which strategic competition is waged, observing that "technology is central to today's geopolitical competition and to the future of our national security, economy and democracy."[2] In defining the U.S. Department of Defense's (DoD's) technology strategy, the Office of the Under Secretary of Defense for Research and Engineering (OUSD[R&E]) recognizes the role of a rising China in motivating prioritization of a set of critical technology areas (CTAs).[3] These geopolitical realities—the existence of state-level strategic competition and the centrality of technology to the outcome of this competition—motivates an analytic task: the development of methodologies for assessing the state of play of technological competition. This report details one such methodology—a DoD-specific net technical assessment (NTA) approach, which, for a given technology area, is designed to characterize the technical state of the art, measure relative national standing, and identify and assess technology applications for strategically important military scenarios.

---

[1] White House, *National Security Strategy*, October 2022, p. 23.

[2] White House, 2022, p. 32.

[3] OUSD(R&E), "USD(R&E) Technology Vision for an Era of Competition," February 1, 2022.

Traditional net assessment, which is described in detail in the next chapter, seeks to connect a nation's ability to execute military operational tasks to achieve higher-level strategic objectives during competition with a competitor. The top four levels of Figure 1.1 illustrate the scope of traditional net assessment.[4] Even when analysts hold technology constant, the complexity of assessing the top four levels often requires analysts to scope the problem to specific mission domains—such as maritime control, air power, ground forces, or nuclear weapons or alternatively to focus on a specific scenario or conflict. The addition of technology as a variable greatly increases the complexity of net assessment.

As shown in Figure 1.1, inclusion of technology adds at least two layers of complexity. In concept, a military technology is enabled by several different subtechnologies. For example, high energy lasers are derived from laser generation technologies, advanced optics, and power and thermal technologies. Inversely, such subtechnologies as advanced computing or artificial intelligence are not specifically military in nature but underpin numerous

FIGURE 1.1
Strategies-to-Technologies Built on Strategies-to-Tasks Framework

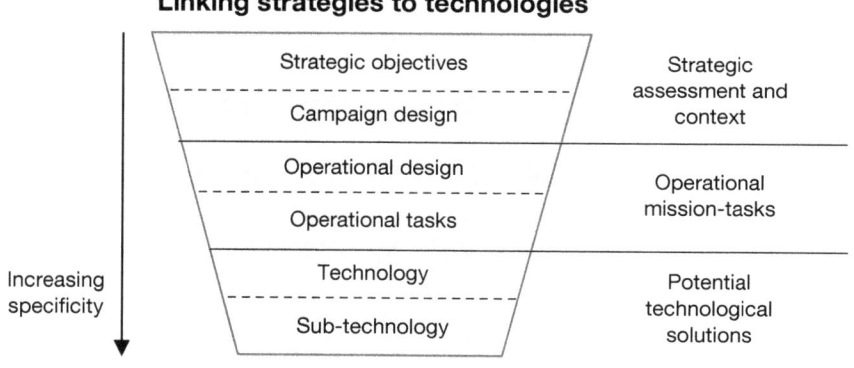

SOURCE: Adapted from Thaler, 1993.

---

[4] David E. Thaler, *Strategies to Tasks: A Framework for Linking Means and Ends*, RAND Corporation, MR-300-AF, 1993.

military technologies, such as advanced radars or cyber defense. Subtechnologies also accelerate the development of other such subtechnologies as advanced materials or biotechnology.

Conducting a DoD NTA that addresses technology requires linking the strategic objectives through campaigns, operations, and technologies via an analytic process that can be executed fast enough to inform ongoing policy decisions. Given the annual budget cycle of DoD, we sought to develop a framework to assess multiple technologies in less than a calendar year.

To this end, this DoD NTA is a multiphased research approach consisting of three major steps: a technical level-setting, a structured analytic workshop, and intelligence community engagement. Each of these steps is designed to provide a critical analytical input into the overall NTA process. For example, via technological horizon scanning, subject-matter elicitation, and bibliometric analysis of the focal technology, the technical level-setting research task establishes the technological state of the art, anticipated progress, and relative national standing. The technical level-setting then informs the structured analytic workshop by establishing among workshop participants a baseline understanding of current technological maturity, applications, and national capability for the focal technological area. Both the technical level-setting and the defense implications from the workshop are then shared with the intelligence community to understand further how competitor activity might affect the assessment. The final step of the NTA methodology is the documentation of results in three artifacts: a technical level-setting report, a workshop summary, and an executive briefing.

During 2021–2022, we executed this process for 14 technologies centered around DoD's CTAs in cooperation with partner federally funded research and development centers (FFRDCs) and university affiliated research centers (UARCs).[5] Three CTAs (space, integrated sensing, and trusted artificial

---

[5] This process was executed in collaboration with government agencies, other FFRDCs, and UARCs. The Office of Strategic Intelligence & Analysis (OSI&A) led the overall coordination process and funded the work. MITRE, Massachusetts Institute of Technology Lincoln Laboratory (MIT LL), Johns Hopkins University Applied Physics Laboratory (JHU APL), Institute for Defense Analyses (IDA), and Georgia Tech Research Institute (GTRI) provided technical expertise, writing, and technology level-setting documents on each CTA. Our role was to assist in the design of the NTA methodology and lead the methodology's execution for 17 distinct technology areas.

intelligence and autonomy) were split into subtopics (shown in parentheses in Table 1.1) due to the wide variety of technologies associated with those CTAs.

The objective of this report is to describe the NTA methodology. The next chapter describes the methodology's intellectual foundations in net assessment, a well-established practice of evaluating the posture of competing countries vis-à-vis their strategic military objectives. Chapter 3 describes

TABLE 1.1

Net Technical Assessment Workshop Schedule 2021–2022

Number	Focus	Tech Lead
1	quantum	MITRE
2	directed energy	MIT LL
3	hypersonics	JHU APL
4	biotechnology	IDA
5	microelectronics	GTRI
6	space	IDA
7	space (space domain awareness)	IDA
8	integrated sensing (cyber and information operations)	MITRE
9	integrated sensing (electronic warfare and signals intelligence)	GTRI
10	integrated network systems of systems	GTRI
11	trusted artificial intelligence and autonomy (artificial intelligence)	GTRI
12	trusted artificial intelligence and autonomy (autonomous systems)	JHU APL
13	human-machine interface	MIT LL
14	future/next generation	MITRE
15	advanced software	JHU APL
16	advanced materials	IDA
17	renewable energy, generation, and storage	IDA

our NTA methodology in detail. Chapter 4 illustrates how the NTA works in practice by tracing how each step in the process contributed to an exemplar analytical result. Chapter 5 concludes with some final thoughts on the utility of NTAs and how DoD can best leverage such activities to maintain critical advantages.

CHAPTER 2

Applying Net Assessment to Technology

This DoD NTA methodology is informed by the long-standing net assessment philosophy championed by DoD's Office of Net Assessment (ONA).[1] *Net assessment* is an approach to evaluating strategic military competition between countries—more of a mission than a specific methodology. Whereas net assessment considers the full set of variables that affects strategic competition, our NTA process uses the principles of the net assessment to evaluate a single dimension of this competition: technology. This section describes the foundational concepts of net assessment and how we leveraged these concepts to develop our NTA methodology.

At first blush, the process of conducting a net assessment for a given technology might appear straightforward. After all, there are established and relatively accurate methods of measuring a country's level of technological sophistication for a given technology.[2] Given that calculation of "net" can mean comparison on some commensurable metric, a naive NTA for a given technology might simply (1) carefully assess the quality of the technical state of the art for a set of competitor countries, (2) compare each country's results, and (3) declare the country with the greatest measured technical sophistication the winner. Although these steps are probably a necessary part of any sound NTA process, the foundational research of Andrew Marshall and the net assessment community demonstrate this naive approach

---

[1]   This chapter was updated in June 2024 to correct instances of *NTA* that were intended to be *net assessment*.

[2]   Comparisons of a technology on a set of specifications (e.g., maximum speed, payload, range) is one approach. Bibliometric-based approaches offer another. Global commercial leadership is another.

to be insufficient to assess the technology balance and its implications on the balance of power.[3]

The practice of net assessment was largely developed by Andrew Marshall during his time at RAND and ONA, an organization he founded and led from 1973–2015.[4] Marshall developed the initial intellectual scaffolding for net assessment while studying Soviet military innovation with Joseph Loftus at RAND in the 1950s.[5] Investigating the Soviet Union's process for acquiring and developing major military systems—they looked specifically at air defense technology, long-range ballistic missiles, and nuclear weapons—convinced Marshall that to correctly understand a country's strategic behavior requires abandoning traditional (rational actor-based) approaches to evaluating military power in favor of a novel approach built on the detailed study of the myriad factors that affect national strategic behavior in practice. Marshall summarizes the aim of net assessment, stating, "the hope is to replace the current rational process model with something better, something that reflects more accurately the context and the constraints within which Soviet military posture incrementally evolves, as the result of a sequence of decisions over many years."[6]

Net assessment, as developed and practiced by Marshall and other researchers at ONA, has at least six defining principles:

1. Military power is scenario-dependent.
2. The nations under assessment are in a competition.
3. Nations, militaries, and their industrial bases are not monolithic, all-knowing, or completely rational.

---

[3] A. W. Marshall, *Problems of Estimating Military Power*, RAND Corporation, P-3417, 1966.

[4] Mie Augier, "Thinking About War and Peace: Andrew Marshall and the Early Development of the Intellectual Foundations for Net Assessment," *Comparative Strategy*, Vol. 32, No. 1, 2013.

[5] Dmitry Adamsky, "The Art of Net Assessment and Uncovering Foreign Military Innovations: Learning from Andrew W. Marshall's Legacy," *Journal of Strategic Studies*, Vol. 43, No. 5, July 22, 2020.

[6] Andrew W. Marshall, "Improving Intelligence Estimates through the Study of Organizational Behavior," paper presented to the RAND Board of Trustees, March 15, 1968, p. 2 (as cited in Augier, 2013).

4. Net assessments are multidisciplinary.
5. Net assessments emphasize diagnostic nature.
6. Competition is considered over some period (i.e., multi-move competition).

We designed our NTA based on these principles. The remainder of this section defines the principles of net assessment and describes how they are incorporated into our NTA methodology.

First, net assessment assumes military power to be based on the practical utility of military resources for a specific operational or competitive context (i.e., military power is assumed to be scenario-dependent).[7] The upshot of this insight is that the like-for-like comparison of a given force element alone is a fundamentally erroneous means of evaluating military resources.[8] Specifically, the naive comparison of like forces embeds a mistaken understanding of the nature of international competition and conflict. In most cases of competition and conflict, military technologies are not mitigated by like technologies, rather they are mitigated via distinct technologies that counteract the effects of the initiating technology (either directly or via deterrence). Offensive missiles are not countered by offensive missiles but rather by missile defense systems. Cyberattacks are not thwarted by more malware but rather by the defender's cybersecurity posture. The naive NTA process described above fails to address the reality that the correct pairing for comparison is often technology A (country X) versus countertechnology B (country Y), not technology A (country X) versus technology A (country Y). Marshall summarizes this point by emphasizing that the value of a country's forces is derived from their utility versus an opponent, stating,

> Most attempts to explicitly measure military power are mere tabulations of forces of various sorts: the number of men under arms, the number of weapons of a given type, etc. This is itself an evasion of the

---

[7] Because military power is scenario-dependent and relative, net assessment does not produce estimates that are transitive. That is, if country A is found to have superior forces to country B, and country B is found to have superior forces to country C, it is not possible to conclude from these facts that country A's forces are superior to those of country C (Marshall, 1966).

[8] Marshall, 1966.

problem of estimating military power since it says nothing about the actual capabilities of the forces of one country to deal with another.[9]

In designing our NTA methodology, we avoided the shortcoming of like-for-like comparison by assessing technologies within the context of a set of scenarios tied to U.S. strategic priorities (see Chapter 3). This approach assured that the focal technology was assessed based on its contribution to operational advantage in a relevant mission setting. Additionally, because these are competition- and conflict-based scenarios, the technologies are always assessed with respect to the relevant competitor countertechnology, both their technical capacity and their anticipated strategy.

Furthermore, by setting the identification of technology opportunities within feasible and high-priority operational settings, we adhered to a body of literature that found military innovation typically commences following the clear articulation of a strategic-level operational problem.[10]

Second, net assessment focuses on behavior among competitors. It is fundamental to net assessment that the major country-level actors under scrutiny are engaged in competition. Competition here simply means that the country-level actors are each aiming to accrue a scarce objective—such as power or influence—via a contested (i.e., noncooperative) process.[11] Final assessment under net assessment is thus relative: A country's position is assessed relative to that of its competitor's. Marshall succinctly summarizes the inherently comparative nature of net assessment, stating, "Estimating the

---

[9] Marshall contends that this logical flaw—which he calls *"symmetry syndrome"*—even negatively affects major defense acquisition decisions, observing, "If an opponent buys bombers, we tend more to increase our bomber forces, rather than to increase our air defenses; when an opponent deploys an ABM [antiballistic missile] system, we deploy an ABM system" (Marshall, 1966, p. 10).

[10] Adam R. Grissom, Caitlin Lee, and Karl P. Mueller, *Innovation in the United States Air Force: Evidence from Six Cases*, RAND Corporation, RR-1207-AF, 2016; Jon Schmid, *The Determinants of Military Technology Innovation and Diffusion*, dissertation, Georgia Institute of Technology, 2018.

[11] For a thorough definition of international competition and discussion of the current period of competition in historical context, see Michael J. Mazarr, Jonathan S. Blake, Abigail Casey, Tim McDonald, Stephanie Pezard, and Michael Spirtas, *Understanding the Emerging Era of International Competition: Theoretical and Historical Perspectives*, RAND Corporation, RR-2726-AF, 2018.

military power of the United States, or any other country, can only be done relative to that of another country, or set of countries viewed as an alliance."[12]

We incorporated this principle into our NTA methodology by designing scenarios around recognized fronts of strategic competition. Although our NTA is fundamentally a comparative process, it does take the perspective of the United States in its overall assessment.[13] That is, our NTA methodology seeks to assess the implications of the technology under scrutiny for executing operations to advance U.S. strategic objectives.[14]

Third, net assessment does not view countries or organizations as *unitary actors*. That is, groups are not treated as a single entity that acts rationally to maximize a stable set of preferences. Instead, net assessment decomposes an organization into subparts based on differences on important dimensions, including interests, power, resources, and culture. It is only when these pieces are understood that anything meaningful about the larger organization can be determined. Marshall draws the contrast between traditional forms of assessment and net assessment by underscoring the former's failure to accurately reflect the real-world behavior of organizations, stating, "Most discussion and forms of analysis tend to treat governments, military organizations, etc., as though they were equivalent to individual rational decisionmakers and not the complicated bureaucratic institutions that they in fact are. We know that decision-making within large organizations and government bureaucracy is not like that which is predicted on the basic of models of rational optimizing behavior."[15]

Net assessment's focus on heterogeneity within organizations stands in contrast to operations research, systems analysis, and neo-realist political theory, which tend to treat countries and organizations as unitary actors.

---

[12] Marshall, 1966, p. 2.

[13] This is not to say that competitor perspectives are not considered. In fact, when a technology is considered within a mission context, the competitor's objectives are overtly taken into account. To say we took the perspective of the United States means we were focused on the utility of the technology (which may be used by the United States and adversaries) to U.S. interests.

[14] The process described here could also be executed by taking the perspective of other countries involved in competition.

[15] Marshall, 1966, p. 10.

Such approaches are well suited to highly structured problems: those for which parameters can be precisely estimated, the universe of determinants is well specified, and uncertainty is low. Game theoretic and computational methods can be applied to such problems to yield precise estimates of unknown parameters.

However, net assessment practitioners tend to view such methods as ill-suited to the problem of strategic political interaction. Strategic competition is viewed by such practitioners as context-dependent, dynamic with respect to time, and characterized by a high degree of uncertainty. In lieu of game theory and other analytic methods, net assessment favors wargaming, scenario-based workshops, and trend analysis. These methods are justified, according to net assessment, not by any *ex ante* assumption to their superiority to analytic methods but rather their fit to assessment of the type of problems on which net assessment focuses. That is, Marshall's rebuke of computational modeling related not to general utility of the method but rather the misalignment of the method to the problem at hand.[16] Bracken describes net assessment's view of the shortcomings of computational modeling, stating, "They [computational models] rarely discuss the intense uncertainties that exist in relationships among the variables, and they make assumptions more to ease the modeling task than to represent actual relationships in the world."[17] The methodology described in this report adheres to Marshall's advocacy of scenario-based analysis when considering the thorny problem of strategic competition. Specifically, we made no assumptions regarding the rationality of the relevant actors and situated the assessment of technology applications in specific scenarios. Diversity of participants in our NTA process—technologists, warfighters, strategists, force planners, and

---

[16] Although rooted in theory, Marshall's motivation to shift from a rational and unitary model of state behavior to incorporating limited rationality and organizational behavior was strictly practical. Marshall and Loftus observed the prevailing approaches for understanding Soviet behavior were inaccurate and yielded inaccurate estimates of future Soviet military posture. The motivation for assessing Soviet behavior in a way that was cognizant of organizational psychology, bounded rationality, and bureaucratic politics was simply to produce better estimates.

[17] Paul Bracken, "Net Assessment: A Practical Guide," *Parameters*, Vol. 36, No. 1, 2006, p. 99.

regional experts—bring different knowledge backgrounds and organizational perspectives into the discussion.

Fourth, net assessment seeks a diversity of perspectives. The tools, concepts, and ideas used to conduct net assessment are found across a wide variety of academic disciplines, including psychology, business strategy, organizational behavior, economics, political science, and history. Interdisciplinarity in net assessment is motivated, in turn, by pragmatism. Tools, ideas, and methods are brought to bear on a problem because they exert analytical leverage, not simply to ensure interdisciplinarity for its own sake. As will be described in greater detail in the next section, our NTA methodology explicitly incorporates a multistep process to ensure a rich information environment that includes multiple perspectives. The team that does the initial technology scoping and level-setting participates in an operational workshop with diverse participants, and the findings are reviewed by the intelligence community before being finalized.

Fifth, net assessment is diagnostic. Net assessment seeks to describe not prescribe. This characteristic owes to Marshall's contention that assessment that seeks to produce policy recommendations can lead researchers to distort the empirical analysis. In the words of Marshall,

> [t]he use of net assessment is intended to be diagnostic. It will highlight efficiency and inefficiency in the way we and others do things, and areas of comparative advantage with respect to our rivals. It is not intended to provide recommendation as to force levels or force structures as an output.[18]

Again, our NTA process adheres to this tenet: The methodology described below stops at analysis. No specific policy questions are posed in the NTA process, and it does not provide DoD with policy recommendations based on this analysis. Although our approach is strictly diagnostic, it does seek to inform DoD processes that allocate resources. In fact, executing the NTA process 17 times over a highly compressed schedule of 18 months

---

[18] Andrew W. Marshall, "The Nature and Scope of Net Assessments," memo for the record, National Security Council, August 16, 1972, p. 2.

sought to ensure the results could inform ongoing decisions regarding the technology priorities of DoD.

Sixth, net assessment is not a static analysis of the state of affairs but rather considers change over time. The typical time horizon for a net assessment is the duration of the competition under scrutiny.[19] Bracken describes the focus on change over time, stating, "One of the greatest contributions of net assessment is that it calls for consciously thinking about the time span of the competition you are in."[20]

The NTA methodology described below incorporates change over time overtly. Technology development from basic research to laboratory demonstrations to operational systems can take decades. Even mature technologies can take years to integrate into existing military platforms given funding, engineering, and training constraints. One of the major dimensions on which each technology opportunity (i.e., a scenario-specific use case for a given technology) is the development and fielding timeline. The NTA process below seeks to explore and understand the timeline constraints on an envisioned military capability and whether it is the maturity of technology, system integration, funding, or other policy factors that drive that timeline.

In addition to adhering to Marshall's principles of sound net assessment, the methodology proposed here seeks to be generalizable to a broad set of technological domains. The desire for generalizability owes largely to the character of the research task for which the methodology was designed: We were asked by DoD to conduct NTAs on the 11 modernization priorities which evolved into 14 CTAs, which exhibit substantial heterogeneity in terms of domain size, research and development players, and civilian application. Thus, the research tasks described in this report can be applied to any technology domain, including such military-specific topics as hypersonic missiles, broad civilian-oriented fields (i.e., biotechnology), warfighting domains (i.e., space with a variety of technologies), and important industrial sectors (i.e., microelectronics).

One assumption and one limitation of the proposed methodology are worth highlighting. An implicit assumption of the assessment of the state of

---

[19] Bracken, 2006.

[20] Bracken, 2006, p. 94.

the art research task (described in Chapter 3) is that a country would prefer to lead its competitors in terms of indigenous technological capability for a given technology. Although this assumption presumably holds in most cases of military-relevant technology, there might be instances in which it is preferable for a country (even a large and technologically sophisticated one) to attain a given technology from abroad. This decision—of whether to develop a technology domestically or import it—is not explicitly considered in our NTA demonstration but should not be ignored by military planners.

A limitation of the proposed approach is its inability to assess the relative utility of technologies from distinct CTAs. For example, if two technologies from distinct CTAs compete for the same operational niche, scoping the NTA process to a single CTA does not allow comparison of these technologies or the identification of potential complementarities. To address relative utility or complementarity would require the assessment to be scoped to multiple technology categories. We explicitly considered such an approach but chose to scope each assessment at the level of the CTA because of the importance of establishing a baseline technical understanding among participants, something that would be infeasible if all 14 CTA were considered simultaneously.

CHAPTER 3

The Net Technical Assessment Research Approach in Detail

Summary of Methodology

Traditional net assessment, as described in Chapter 2, does not limit focus to a given technology area. In fact, traditional net assessment is, in part, designed to determine what military technologies might be of utility in a future scenario. However, given the high variance in the rate of technological advancement across technologies, this approach could introduce a large amount of forecasting error.

To make our NTA process manageable, we reversed the traditional process. We started by scoping to a particular technology area that is of particular interest to DoD and U.S. adversaries. We then drew strategic contexts from existing defense strategies and operational plans. We explored the military capabilities enabled by those technologies that are most applicable to the chosen strategic contexts. Finally, we confirmed that our perspectives included the competitors' views on strategy and technology and the resulting competition.

The overall NTA research approach consists of the following three major tasks:

1. scoping technologies and associated trends
2. exploring military operational implications
3. reflecting on competitor perspectives.

This process (summarized in Table 3.1) is executed for each CTA with an initial technical level-setting review, a military structured analyti-

TABLE 3.1

Summary of Net Technical Assessment Methodology

Research Tasks: Approach	Objectives	Deliverables
Technology trends task: technical level-setting	• Scope a set of technologies • Assess technologies' technical state of the art and trends • Capture relative national standing for focal technologies	• Technical level-setting report • Technical level-setting brief • Relative national standing brief
Operational implications task: structured analytic workshop	• Identify and refine scenario-specific technology opportunities for focal technologies • Assess operational advantage and fielding timeline for each technology opportunity	• Technology opportunity list assessed for operational advantage and fielding timeline • Workshop notes (e.g., description of discussion, notable points of expert divergence and convergence)
Competitor perspective task: intelligence community engagement	• Refine all prior assessments based on intelligence	• Refined assessments of technical level-setting, national standing, and technology opportunities
Summary of full NTA process	• Evaluation of the expected impact of a technology on strategic competition	• Final NTA report for each CTA

cal workshop, and intelligence community engagement. During technical level-setting, researchers scope the technologies in play and investigate the technical state of the art and relative national performance for the target technologies. This task is performed in the months preceding the structured analytic workshop for the CTAs under assessment. The findings of the technical level-setting process are presented during the workshop to ensure participants begin with a shared technical understanding of the focal technologies.

The structured analytic workshop is a daylong in-person event meant to identify how the focal technologies might be used in distinct scenarios to achieve specific operational objectives. The principal output of the structured analytic workshop is a set of technology opportunities that have been evaluated by workshop participants on two dimensions: operational advantage and fielding timeline. Given the nature of a one-day workshop, the findings are not definitive but provide insight into priorities of the defense community in those contexts and sources of uncertainty about the advantages and timelines for those technology opportunities.

The final step of the NTA research approach is intelligence community engagement. For some technologies, the intelligence community is privy to information about the technical state of the art that might modify the assessments made during tasks 1 and 2. If the intelligence community is aware of technical advancements, barriers, or applications that are not reflected in the preliminary findings, adjustments to the assessment of each technology will be made to the final NTA documentation.

Technical Level-Setting

Scoping Net Technical Assessment

The NTA process is conducted at the level of the CTA.[1] The list below presents the 14 CTAs for which we conducted NTAs. However, to accurately assess relevant dimensions of a technology (i.e., its maturity or operational impact) requires disaggregating the CTA into its component technologies.

---

[1] DoD critical technology areas are a set of technology areas deemed to be of particular import and high investment priority by DoD.

For example, assessment of operational impact for a given technology (e.g., quantum key distribution, atomic clocks, Rydberg sensors) can be conducted by considering the operational advantage afforded by that technology across a set of scenarios. Because CTAs consist of dozens if not hundreds of technologies, assessment of the operational impact of a CTA, such as quantum science, is impossible without first assessing the impact of its component technologies. The first analytic step in the NTA process is to decompose each CTA into a set of technologies. For each CTA, a team of technical experts selects three to a dozen technologies determined to be technically realizable in the medium term (less than 20 years) and have high military impact.[2]

This down-select process involves researcher discretion, creating the possibility that important technologies could be left out of the analysis. To mitigate this threat, we took two steps. First, the initial technology selection process was conducted by a team of subject-matter experts with deep expertise in the CTA and the operational realties of major U.S. military missions. When possible, the candidate technologies were reviewed by the OUSD(R&E) principal director for that CTA. Second, during the NTA workshops and intelligence community engagements, we allowed participants (made up of a mix of technical experts and operators) to suggest technologies from within the CTA that were not included in the initial level-setting brief. As appropriate, the technology level-setting team could include those technologies in their report, and the summary of the workshop notes would capture the participants' interest in them.

We conducted NTAs for the following DoD CTAs as part of this research:

- advanced computing and software
- advanced materials
- biotechnology
- directed energy
- future generation wireless technology (FutureG)
- human-machine interfaces
- hypersonics

---

[2] These teams typically consisted of Ph.D.-level technical experts in the focal field with experience in considering the military applications of the technologies in question.

- integrated network systems-of-systems
- integrating sensing and cyber
- microelectronics
- quantum science
- renewable energy generation and storage
- space technology
- trusted artificial intelligence and autonomy.[3]

Assessment of the State of the Art (Task 1.1)

In the weeks prior to the NTA workshop, each technology and the technology elements from which it is comprised is assessed on such dimensions as current technical maturity, the rate of technological change, and potential bottlenecks to development. This assessment is documented in a brief and report, which are presented and provided, respectively, to the NTA-structured analytic workshop participants. This research task achieves at least two purposes. First, by documenting the state of the art for each selected technology, task 1.1 creates a technical document highlighting critical technology elements (i.e., the precursors to integrated systems) to the focal technologies—a document that can be used to inform DoD investment decisions. Second, presenting the findings of this task to workshop participants allows participants of diverse backgrounds to begin with a reasonably high-level shared understanding of the technologies and their capabilities. In practice, this task was executed by experts from other FFRDCs and UARCs. Specifically, MITRE, MIT LL, JHU APL, IDA, and GTRI led the multiple assessments of the state of the art for the 17 technology area workshops for which an NTA was conducted.

Assessment of Country-Level Activity and Capacity (Task 1.2)

The second step of the technical level-setting process is to quantitatively assess country-level technological standing for each of the technologies specified in task 1.1. To this end, we began by computing a set of four met-

---

3 OUSD(R&E), 2022.

rics using custom-build patent and publication datasets for a fixed set of countries.

To compute the four metrics requires building patent and scientific publication datasets for each of the technologies to be assessed. To this end, we employed a keyword-based search strategy whereby a set of keywords are used to query patent and publication databases to produce a set of records that are representative of patenting and publishing activity within the technological domain in question. To define a final set of keywords for each technology, we began with a few hundred candidate keywords for each technology and asked a subject-matter expert to use this list to select a final set of search terms by following a "minimize false positives" approach to term selection. That is, a subject-matter expert was asked to select terms that would yield only search results within the target technology scope.[4] As necessary, we engaged with technology subject-matter experts from MITRE, GTRI, IDA, MIT LL, and JHU APL to define a list of terms. In the case of patents, we supplemented the keyword-based search approach by adding patent classification codes to the database search.[5]

These terms are used to construct a patent and publication dataset for each technology. Specifically, the terms and patent classification codes are used to build a Boolean search string to query the Web of Science and Derwent databases. For each technology, these search results were downloaded and used to compute four metrics: high-impact publications, quality-adjusted patents, organizational capacity (i.e., organizations active in the field during the analysis period as either publishers or patent assignees), and collaboration network density. These metrics were selected because they respectively proxy for four important dimensions of national innovation preeminence: scientific output, technological output, organizational capac-

---

[4] Consider, for example, an attempt to define terms for the target technology quantum key distribution. The term *quantum* would produce many relevant quantum key distribution publications and patents. However, this item would also yield many *false positives*: returned results that are outside the target technology area. Thus, the term *quantum* would not be selected to define quantum key distribution.

[5] These codes are assigned to all patents by patent examiners based on technical content of the focal patent.

ity, and the health of the relevant scientific network.[6] These measures provide a standardized estimate of activity and institutional capacity that can be applied to many different technology fields of different scales.[7]

In addition to the four metrics presented in Figure 3.1, for each technology, we conducted additional analysis on the character of domestic research and development ecosystems. For example, analysis of coauthorship networks was used to identify the most important organizations within a given domestic research ecosystem and the patterns of collaboration across distinct entity types. Figure 3.2 presents the coauthorship network for the top publishing organizations in the field of quantum communications. Table 3.2 provides network statistics at the country level and the most important domestic organizations in terms of network degree and network centrality.

One important limitation of these metrics is that they rely on open-source data and thus do not measure a country's scientific or technological progress that has been maintained as state or trade secrets. Task 3 of our research approach (intelligence community engagement) seeks to address this limitation by incorporating classified material into our assessment. In addition to calculating the four metrics for each technology, we conducted research on the relevant industrial base and government initiatives. The output of the assessment of country-level activity and capacity process was documented in a brief, which was presented to participants during the NTA workshop. This analysis is also summarized in the final NTA report for each CTA.

---

[6] Jon Schmid, *An Open-Source Method for Assessing National Scientific and Technological Standing: With Applications to Artificial Intelligence and Machine Learning*, RAND Corporation, RR-A1482-3, 2021.

[7] Where possible, we adjusted count-based metrics to account for inter-country differences in patent and publication quality. Quality-adjusted patents adjust raw patent counts by applying a discount factor based on patent family size, which is correlated with patent quality and the monetary value of the patent (Dietmar Harhoff, Frederic M. Scherer, and Katrin Vopel, "Citations, Family Size, Opposition and the Value of Patent Rights," *Research Policy*, Vol. 32, No. 8, September 2003). To account for variability in publication quality, the high-impact publication metric counts only a country's publications that fall within the top 10 percent of the annual citation distribution.

FIGURE 3.1

Examples Slide: Assessment of Country-Level Activity and Capacity from Quantum Science Net Technical Assessment

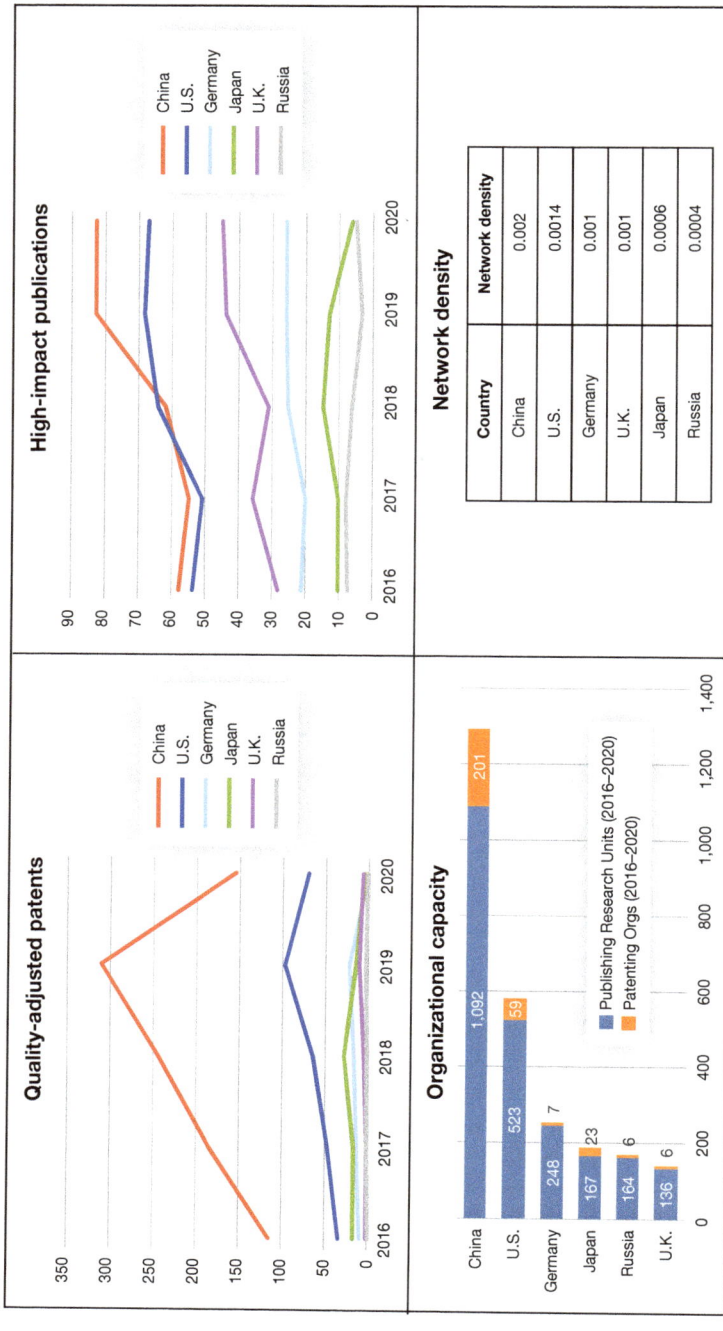

FIGURE 3.2

Examples Visualization: Coauthorship Network, Quantum Communications, 2016–2022

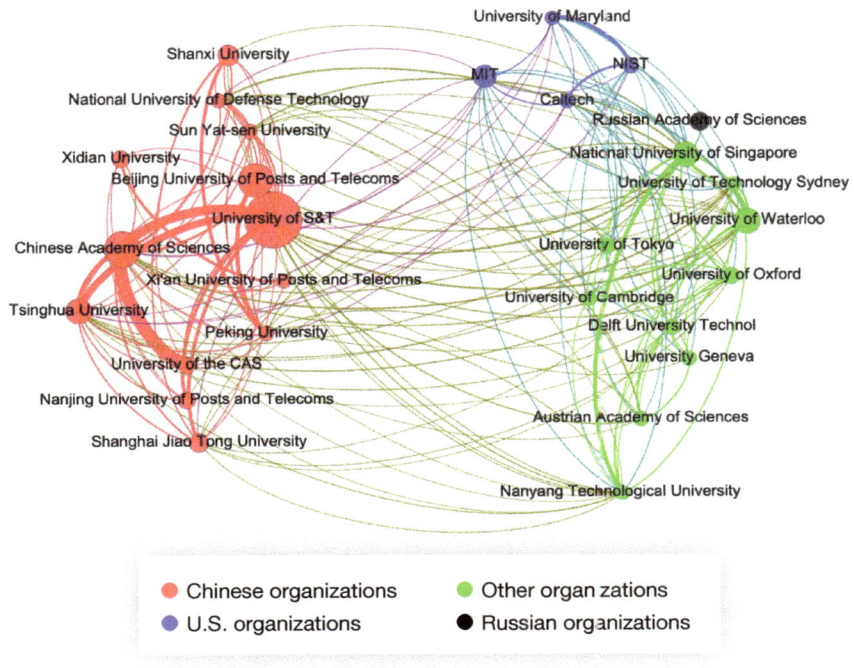

SOURCE: Produced using Clarivate Analytics, "Web of Science Platform," database, undated; author analysis.

Structured Analytic Workshop

For each CTA, we conducted a structured analytic workshop. These in-person events took place over a full day and were split into four primary activities: facilitated technical exchange, presentation of strategic contexts, technology opportunity identification, and a hotwash.[8]

---

[8] A *hotwash* is a debriefing session conducted immediately after a workshop during which participants discuss the overall content, outcomes, and experiences to identify areas for improvement and enable organizers to gather previously unstated information.

TABLE 3.2

Network Metrics for Select Countries, Coauthorship Network, Quantum Communications, 2016–2022

Country	Network Density[a]	Organization with Most Collaborations (Degree)[b]	Most Central Organization (Eigenvector Centrality)
China	0.0020	University of Science and Technology (530)	University of Science and Technology (0.84)
United States	0.0014	Massachusetts Institute of Technology (371)	MIT (0.81)
Germany	0.0010	Max Planck Institute (186)	Max Planck Institute (0.70)
United Kingdom	0.0010	University of Oxford (260)	University of Oxford (0.86)
Japan	0.0006	University of Tokyo (213)	University of Tokyo (0.46)
Russia	0.0004	Russian Quantum Center (221)	Russian Quantum Center (0.58)
South Korea	0.0002	Korea Advanced Institute or Science and Technology (52)	Seoul National University (0.08)

SOURCE: Authors' analysis of Web of Science data.

NOTE: Interestingly, although Canada trails the United States and China in terms of aggregate publication output, the University of Waterloo is the most central organization in the quantum communications network.

[a] Network density of country-specific network (i.e., the subset of publications with at least one author from the focal country).

[b] Degree and centrality computed for the global quantum communications network.

Facilitated Technical Exchange (Task 2.1)

The first part of the structured analytic workshop was designed to establish a common understanding of technologies to be assessed about how the technologies work, potential applications, technical maturity, and country-level capacity and activity. To this end, the structured analytic workshop begins with two presentations. The first presentation is based on the output of the technical level-setting process and focuses on the technologies to be considered, their scientific and technical underpinnings, barriers to development, and potential civilian and military applications.[9] The second is focused on

---

[9] At this point in the workshop, discussion of applications is agnostic with regard to mission. The distinction between a technology application presented during the facili-

relative national standing and consists of the four quantitative metrics for each technology, additional bibliometric analysis, and information on relevant industrial base and government initiatives.

Presentation of Strategic Contexts (Task 2.2)

The overall purpose of strategic context scenarios in the NTA workshops is to provide workshop participants notional but plausible settings for proposing and debating the potential operational applications and advantages of technology opportunities. In assessing the operational utility of a particular technology, one can adopt either of two basic approaches. The first is to start with the technology and then look for scenarios that best fit the technology. The second is to start with scenarios that are reflective of strategic priorities and refrain from taking *ex ante* assumptions about the role of the CTA to be assessed. For the NTA workshops, the latter approach was employed: We selected scenarios based on their relevance to U.S. strategic priorities and were largely agnostic of the utility of the technology under scrutiny.[10]

For each NTA workshop, we presented three strategic contexts. Each scenario consists of a broad strategic setting and a more concrete set of U.S. operational problems or objectives. But even these operational objectives remain intentionally broad (e.g., enable geographically distributed air operations, protect U.S. forces, preserve battlespace awareness). The aim of the strategic contexts is to avoid driving workshop participants to a specific, predetermined outcome and instead open the door to a broad set of potential applications for any of the considered technologies. We sought to capture a wide variety of potential advantages and include niche applications of a given technology within this set of scenarios by incorporating significant within-scenario and cross-scenario variation along multiple dimensions. These dimensions include geographic theater of operations, warfighting domain (land, air, maritime, space, cyber), level of conflict (steady-state

---

tated technical exchange and a technology opportunity is that a *technology opportunity* (as described in task 2.3) is articulated within the context of a scenario and typically involves the definition of a platform and concept of employment.

[10] This approach is based on RAND's strategies to task methodology, a process for linking low-level decisions (e.g., operational tasks or technologies) to high-level strategies (e.g., the objectives of the National Defense Strategy) (Thaler, 1993).

competition, gray zone, major combat operations), and joint warfighting function (intelligence, movement and maneuver, fires, information, protection, sustainment, command and control [C2]).

In building the slate of scenarios for each workshop, we drew from DoD-, joint-, and service-specific strategic documents, along with RAND's ongoing dialogue with senior DoD leaders and experience in DoD wargames. Although the scenarios were tailored to the workshop to ensure a minimal degree of operational relevance for the focal CTA, during each workshop, one scenario focused on steady-state great-power competition, one on gray-zone activity aimed at harming a U.S. partner, and one on major combat operations. We were careful to ensure that each set of scenarios and associated operational objectives captured the seven joint warfighting functions of intelligence, movement and maneuver, fires, information, protection, sustainment, and C2.

Identify Technology Opportunities (Task 2.3)

Following the presentation of each strategic context, participants were asked to nominate technology opportunities. A *technology opportunity* is an application or use case of a technology that is linked to a mission objective within a scenario. For each nomination, participants were asked to briefly describe the technology to be used, its basic concept of employment, and the operational challenge within the scenario that the technology would address. We explicitly encouraged workshop participants to consider ways in which a given technology might provide direct or indirect advantage in a scenario. Even if the application is not direct or immediately obvious, the technology might still offer significant value through such second- and even third-order effects as cost savings, better training, or greater force readiness.

Participants were then asked to assign scores to each technology opportunity on two dimensions: operational impact and the development fielding timeline. Table 3.3 provides the voting scale. Votes were recorded by the research team.

The distribution of scores varied by technology. Figure 3.3 presents the results from a technology for which respondents' scores were highly concentrated in the top right quadrant (i.e., high operational advantage and short development and fielding timelines). Figure 3.4 presents the results from

TABLE 3.3

Voting Scale for Operational Advantage and Development and Fielding Timeline

Score	Operational Advantage	Development and Fielding Timeline
0	None—not operationally relevant	Never—fundamentally physically impossible
1	Limited—operationally relevant but no direct impact	Uncertain—unlikely in the next 20 years
2	Convenient—operationally relevant; some convenient improvements to a military mission	Long term—possible in 10–20 years to demonstrate and integrate or develop
3	Beneficial—operationally advantageous; offers some advantages for a military mission	Medium term—some demonstrations; 5–10 years to integrate or develop
4	Impactful—operationally essential; provides critical advantages for operations	Near term—within future year defense program path for insertion to platforms or prototypes that exist
5	Revolutionary—fundamentally alters how the U.S. military operates	Immediate—can be developed and fielded in 1–2 years.

a technology for which respondents' scores were highly varied. In these graphs, the points represent the mean score for a given technology by participants. Along with documenting participant opinion about operational impact and fielding timelines, the voting process sought to identify technology opportunities with high vote ranges (i.e., participant disagreement). Technology opportunities with the highest vote disparities were selected for further discussion. During this discussion, participants were given an opportunity to explain their votes or clarify the assumptions underlying their assessments. Following this discussion, all participants revoted on technology opportunities with high vote ranges to check for convergence or continued divergence of opinion. This step—focusing on areas of vote divergence, allowing the articulation of strong positions, and revoting—takes inspiration from the Delphi method, a common means of synthesizing individual expert judgments on complex topics.[11]

---

[11] Norman Dalkey and Olaf Helmer, "An Experimental Application of the Delphi Method to the Use of Experts," *Management Science*, Vol. 9, No. 3, April 1963.

FIGURE 3.3

Operational Advantage by Timeline, Exemplar Concentrated Results

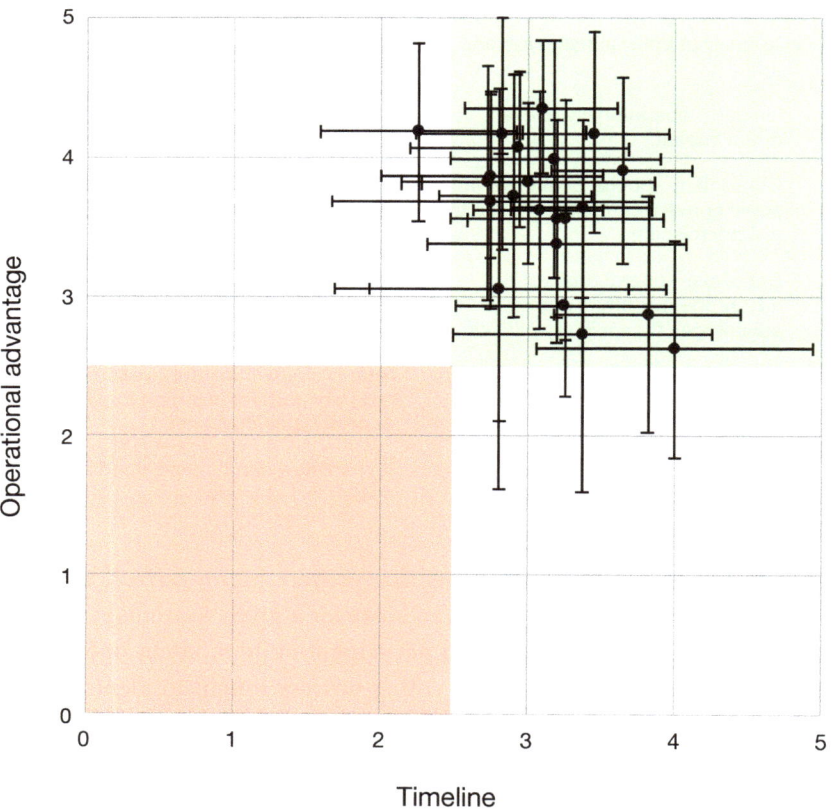

NOTE: Points represent the mean voting score for a given technology by all participants. Lines are standard deviation, indicating different views about operational and technical issues.

Hotwash (Task 2.4)

The hotwash at the end of the day was focused on three elements. First, we allocated time to identify technology areas that were not discussed but are nevertheless relevant to military applications. Typically, this meant addressing the limitations of the initial technology-scoping decision (i.e.,

FIGURE 3.4

Operational Advantage by Timeline, Exemplar High-Variance Results

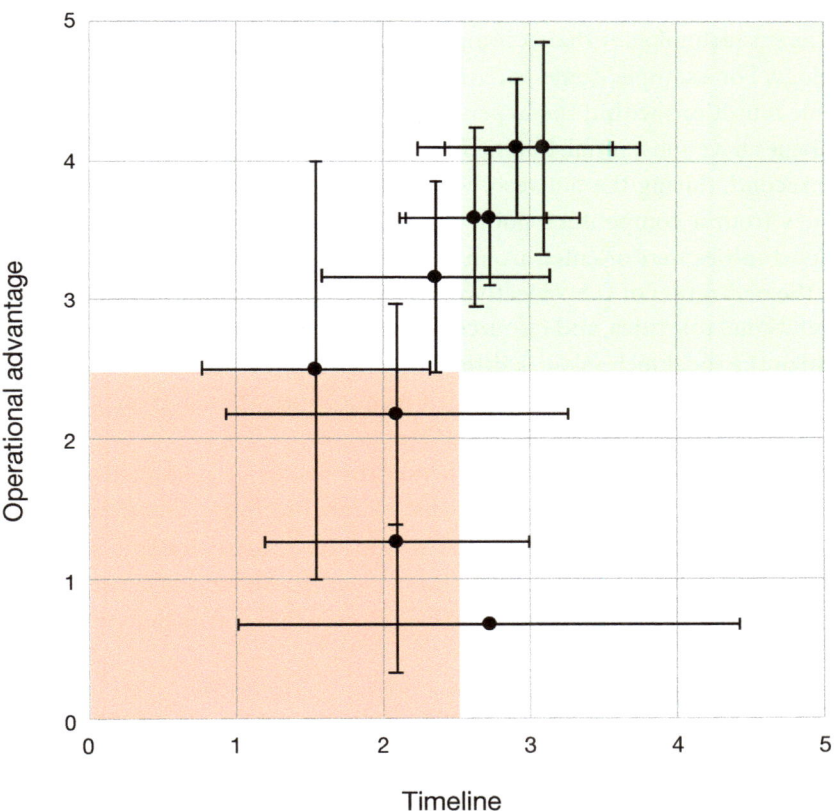

NOTE: Points represent the mean voting score for a given technology by all participants. Lines are standard deviation, indicating different views about operational and technical issues.

determining what portions of very large technology—space technology, biotechnology—would be considered within scope for the NTA workshop). During this portion of the hotwash, we sought to elicit information about what additional technologies participants would have liked to have been included in the workshop. We observed two common variants of this discussion. In the first, participants sought the inclusion of technologies that were included during another NTA workshop (e.g., artificial intelligence can enable many military technologies but was explicitly addressed during the trusted artificial intelligence and autonomy NTA workshop). Second, participants frequently expressed that they would have liked to have included different technologies that accomplish the same mission as the focal technology. For example, some participants expressed a desire to include subsonic munitions within the hypersonic NTA workshop because these technologies have substantial, but incomplete, mission overlap.

Second, during the hotwash, we asked the participants to consider the issues from a competitor's point of view. The workshop and technology opportunities were oriented around a U.S. perspective and articulated based on the realization of U.S. objectives. However, given distinct strategic goals, operational priorities, and resource constraints, adversaries are likely to prioritize the focal technologies differently. During the hotwash, we tried to elicit participant insight about which technologies offered greater relative advantage for Red than for Blue, and vice versa.

Third, we asked participants whether the workshop presented any surprises related to technical level-setting, national technical standing, or the technology opportunities identified within the chosen strategic contexts. During this discussion, we sought also to identify any areas of consensus, or persistent disagreement, that should be captured in the workshop summary.

Intelligence Community Engagement

The final NTA research task is intelligence community engagement. The purpose of this task is to incorporate information that is not available via open sources. For example, RAND analysts in the 1950s were not privy to closely guarded intelligence that Soviet Bison and Bear bombers had significant engine reliability problems. This led to overestimation of Soviet nuclear strike capabilities, which, in turn, affected U.S. military policy.[12] The intelligence community engagement portion of the NTA process sought to avoid such miscalculations by ensuring the integration of information located within distinct organizations.

---

[12] Bracken, 2006.

The intelligence community engagement task centered around a two-hour structured video conference review of draft reports that took place several weeks following the structured analytic workshop. Conducting the intelligence community engagement following the structured analytic workshop allowed us to document the findings of the workshop and present them for review by the intelligence community. Prior to the video conference, the intelligence community was sent the technical level-setting report, a draft NTA summary report, and a three-page read-ahead containing a catalog of technology-specific questions and topics of discussion. The focus of the call was on the preliminary NTA findings. The intelligence community was asked to provide any information that would require us to amend or clarify these findings. Particular attention was paid to the state of competitor technological advancement and the existence or the planned acquisition by adversaries of counters to the focal technologies. In other words, engagement with the intelligence community sought to avoid *mirror imaging* or the tendency to expect the subject of analysis to act and think the way that we do. Notes from this discussion, along with sources provided by the intelligence community, were used to amend the NTA artifacts.

CHAPTER 4

Illustration of the Net Technical Assessment Process

To illustrate how each of these methodological processes contribute to the conclusions drawn in the final assessment, it is useful to focus on a single conclusion from the quantum science NTA report. The quantum science NTA concluded, *inter alia*, that China is the clear global leader in quantum key distribution, but this technology has limited near- or medium-term operational utility that affects U.S. strategic priorities. We can speak freely on these illustrative findings because they are already a matter of public record.[1]

The first analytical input to this conclusion was to divide the CTA into a set of technologies on which the NTA process would focus. In this case, during the technology level-setting process, quantum science was split into three technologies: quantum computing, quantum communications, and quantum sensors. Post-quantum cryptography was also mentioned as a tangential but relevant technology. Quantum communications included quantum key distribution (QKD) and quantum networks. Quantum sensing included a variety of sensors, but a limited number were detailed—such as Rydberg electric field sensors, quantum inertial sensors, and atomic clocks. In summary, QKD is a method of using the properties of quantum information science to share cryptographic keys with near-zero probability of

---

[1] The claim that quantum communications is of limited national security utility has been made elsewhere. For example, the National Security Agency states that it "does not support the usage of QKD or QC to protect communications in National Security Systems" (National Security Agency, "Quantum Key Distribution (QKD) and Quantum Cryptography (QC)," webpage, undated).

eavesdropping; any keys intercepted by a third party would be detected and discarded.

During the assessment of country-level activity and capacity, it was determined that China is the clear global leader in QKD. This finding has been documented in various other sources.[2] Additional evidence of China's position of leadership was provided during this task by observing that over the past few years, China has successfully completed a satellite-to-ground QKD network and QKD over fiber optic cable.[3] In sum, the technology level-setting process defined QKD and identified China as the global leader in QKD. However, assessment of the impact of QKD on U.S. strategic priorities requires considering the technology within specific scenarios. This was achieved during the structured analytic workshop.

During the quantum science NTA structured analytic workshop, three scenarios were used that covered the full conflict continuum, from competition to overt conflict. The QKD-related technology opportunities for these scenarios related to enabling secure communications between friendly platforms. However, it was determined that the operational advantage provided by these technology opportunities was low.[4] For one, it was observed that QKD only protects against signal interception, not signal degrada-

---

[2] Edward Parker, Daniel Gonzales, Ajay K. Kochhar, Sydney Litterer, Kathryn O'Connor, Jon Schmid, Keller Scholl, Richard Silberglitt, Joan Chang, Christopher A. Eusebi, and Scott W. Harold, *An Assessment of the U.S. and Chinese Industrial Bases in Quantum Technology*, RAND Corporation, RR-A869-1, 2022; Tom Stefanick, "The State of U.S.-China Quantum Data Security Competition," Brookings Institute, September 18, 2020.

[3] Yu-Ao Chen, Qiang Zhang, Teng-Yun Chen, Wen-Qi Cai, Sheng-Kai Liao, Jun Zhang, Kai Chen, Juan Yin, Ji-Gang Ren, Zhu Chen, Sheng-Long Han, Qing Yu, Ken Liang, Fei Zhou, Xiao Yuan, Mei-Sheng Zhao, Tian-Yin Wang, Xiao Jiang, Liang Zhang, Wei-Yue Liu, Yang Li, Qi Shen, Yuan Cao, Chao-Yang Lu, Rong Shu, Jian-Yu Wang, Li Li, Nai-Le Liu, Feihu Xu, Xiang-Bin Wang, Cheng-Zhi Peng, and Jian-Wei Pan, "An Integrated Space-to-Ground Quantum Communication Network over 4,600 Kilometers," *Nature*, Vol. 589, January 6, 2021.

[4] This assessment is consistent with that of the Defense Science Board that concluded, "QKD has not been implemented with sufficient capability or security to be deployed for DoD mission use" (Defense Science Board, *Applications of Quantum Technologies: Executive Summary*, Office of the Under Secretary of Defense for Research and Engineering, October 2019, p. 4).

tion (a major challenge in many high priority missions). Furthermore, it was observed that viable alternatives—such as establishing symmetric keys between friendly platforms—had favorable relative technical qualities to QKD-based approaches. Thus, QKD-related technology opportunities were found to yield little operational advantage to friendly or competitor forces in the scenarios considered.

Thus, the composite conclusion—China is the clear global leader in QKD, but this technology has limited near- or medium-term operational utility that affects U.S. strategic priorities—is derived from distinct phases of the NTA process. The first part of the conclusion (i.e., China's leadership) is based on a thorough assessment of the scientific and technological landscape. The latter (i.e., QKD's limited operational utility) is based on the limited expected utility across examined scenarios.

This chapter selected a single finding to illustrate how the elements of the methodology yield insight about a given CTA. It is worth noting that the character of insight gleaned was broad. Other insights pertained to the importance of the domestic industrial base for certain technology areas; the dependence of certain technologies on critical (sometimes shared) precursor scientific or technological breakthroughs; the importance of such ancillary engineering advancements as those that yield improvement in size, weight, and power; the high variability of cost associated with realizing progress across considered technologies; and the importance of a wide variety of nontechnical factors (e.g., the doctrine, organization, training, materiel, leadership and education, personnel, facilities, and policy factors) that can be expected to influence fielding and deployment timelines.

CHAPTER 5

Conclusion

This report lays out a process for conducting NTA that is generalizable and consistent with Marshall's principles of net assessment. Regarding the former, execution of the methodology on a set of highly heterogeneous technological domains provides confidence in the methodology's generalizability. However, because an NTA depends on assessment of the state of the art and the contemporary security environment, both concepts that are intrinsically time dependent, the findings of any NTA process are likely only applicable to the period during which these variables hold true. Put simply, the output of an NTA is not the final word on a topic. It would therefore be prudent to incorporate into this methodology an updating process whereby the inputs of the NTA process are checked against the status quo and reassessed when necessary.

Regarding net assessment, design elements of the methodology sought to ensure consistency with Marshall's tenets of net assessment. To prevent a naive like-for-like comparison of competitors' realized technological maturity or reliance on unrealistic assumptions of actor rationality or unitary state actors, we situated the identification and assessment of technology opportunities within a set of high-priority operational scenarios involving competitors. To ensure heterogeneity in academic disciplines and perspective, we borrowed ideas from a variety of academic disciplines and drew on participants with diverse professional experiences.

Abbreviations

C2	command and control
CTA	critical technology area
DoD	U.S. Department of Defense
FFRDC	federally funded research and development center
GTRI	Georgia Tech Research Institute
IDA	Institute for Defense Analyses
JHU APL	Johns Hopkins University Applied Physics Laboratory
MIT	Massachusetts Institute of Technology
MIT LL	MIT Lincoln Laboratory
NTA	net technical assessment
ONA	Office of Net Assessment
OUSD(R&E)	Office of the Under Secretary of Defense for Research and Engineering
QKD	quantum key distribution
UARC	university affiliated research center

References

Adamsky, Dmitry, "The Art of Net Assessment and Uncovering Foreign Military Innovations: Learning from Andrew W. Marshall's Legacy," *Journal of Strategic Studies*, Vol. 43, No. 5, July 22, 2020.

Augier, Mie, "Thinking About War and Peace: Andrew Marshall and the Early Development of the Intellectual Foundations for Net Assessment," *Comparative Strategy*, Vol. 32, No. 1, 2013.

Bracken, Paul, "Net Assessment: A Practical Guide," *Parameters*, Vol. 36, No. 1, 2006.

Chen, Yu-Ao, Qiang Zhang, Teng-Yun Chen, Wen-Qi Cai, Sheng-Kai Liao, Jun Zhang, Kai Chen, Juan Yin, Ji-Gang Ren, Zhu Chen, Sheng-Long Han, Qing Yu, Ken Liang, Fei Zhou, Xiao Yuan, Mei-Sheng Zhao, Tian-Yin Wang, Xiao Jiang, Liang Zhang, Wei-Yue Liu, Yang Li, Qi Shen, Yuan Cao, Chao-Yang Lu, Rong Shu, Jian-Yu Wang, Li Li, Nai-Le Liu, Feihu Xu, Xiang-Bin Wang, Cheng-Zhi Peng, and Jian-Wei Pan, "An Integrated Space-to-Ground Quantum Communication Network over 4,600 Kilometers," *Nature*, Vol. 589, January 6, 2021.

Clarivate Analytics, "Web of Science Platform," database, undated. As of January 30, 2024:
https://clarivate.com/webofsciencegroup/solutions/web-of-science

Dalkey, Norman, and Olaf Helmer, "An Experimental Application of the Delphi Method to the Use of Experts," *Management Science*, Vol. 9, No. 3, April 1963.

Defense Science Board, *Applications of Quantum Technologies: Executive Summary*, Office of the Under Secretary of Defense for Research and Engineering, October 2019.

Grissom, Adam R., Caitlin Lee, and Karl P. Mueller, *Innovation in the United States Air Force: Evidence from Six Cases*, RAND Corporation, RR-1207-AF, 2016. As of January 29, 2024:
https://www.rand.org/pubs/research_reports/RR1207.html

Harhoff, Dietmar, Frederic M. Scherer, and Katrin Vopel, "Citations, Family Size, Opposition and the Value of Patent Rights," *Research Policy*, Vol. 32, No. 8, September 2003.

Marshall, A. W., "Problems of Estimating Military Power," RAND Corporation, P-3417, 1966. As of March 20, 2023:
https://www.rand.org/pubs/papers/P3417.html

Marshall, Andrew W., "Improving Intelligence Estimates through the Study of Organizational Behavior," paper presented to the RAND Board of Trustees, March 15, 1968.

Marshall, Andrew W., "The Nature and Scope of Net Assessments," memo for the record, National Security Council, August 16, 1972.

Mazarr, Michael J., Jonathan S. Blake, Abigail Casey, Tim McDonald, Stephanie Pezard, and Michael Spirtas, *Understanding the Emerging Era of International Competition: Theoretical and Historical Perspectives*, RAND Corporation, RR-2726-AF, 2018. As of March 21, 2023: https://www.rand.org/pubs/research_reports/RR2726.html

National Security Agency, "Quantum Key Distribution (QKD) and Quantum Cryptography (QC)," webpage, undated. As of January 30, 2024: https://www.nsa.gov/Cybersecurity/ Quantum-Key-Distribution-QKD-and-Quantum-Cryptography-QC/

Office of the Under Secretary of Defense for Research and Engineering, "USD(R&E) Technology Vision for an Era of Competition," February 1, 2022.

OUSD(R&E)—*See* Office of the Under Secretary of Defense for Research and Engineering.

Parker, Edward, Daniel Gonzales, Ajay K. Kochhar, Sydney Litterer, Kathryn O'Connor, Jon Schmid, Keller Scholl, Richard Silberglitt, Joan Chang, Christopher A. Eusebi, and Scott W. Harold, *An Assessment of the U.S. and Chinese Industrial Bases in Quantum Technology*, RAND Corporation, RR-A869-1, 2022. As of January 30, 2024: https://www.rand.org/pubs/research_reports/RRA869-1.html

Schmid, Jon, *The Determinants of Military Technology Innovation and Diffusion*, dissertation, Georgia Institute of Technology, 2018.

Schmid, Jon, *An Open-Source Method for Assessing National Scientific and Technological Standing: With Applications to Artificial Intelligence and Machine Learning*, RAND Corporation, RR-A1482-3, 2021. As of March 20, 2023: https://www.rand.org/pubs/research_reports/RRA1482-3.html

Stefanick, Tom, "The State of U.S.-China Quantum Data Security Competition," Brookings Institute, September 18, 2020.

Thaler, David E., *Strategies to Tasks: A Framework for Linking Means and Ends*, RAND Corporation, MR-300-AF, 1993. As of March 21, 2023: https://www.rand.org/pubs/monograph_reports/MR300.html

White House, *National Security Strategy*, October 2022.

Milton Keynes UK
Ingram Content Group UK Ltd.
UKHW021352231024
450026UK00007B/531